AF700916

THIS BOOK BELONGS TO

____

____

A MATHEMATICIAN

Embrobi LLC

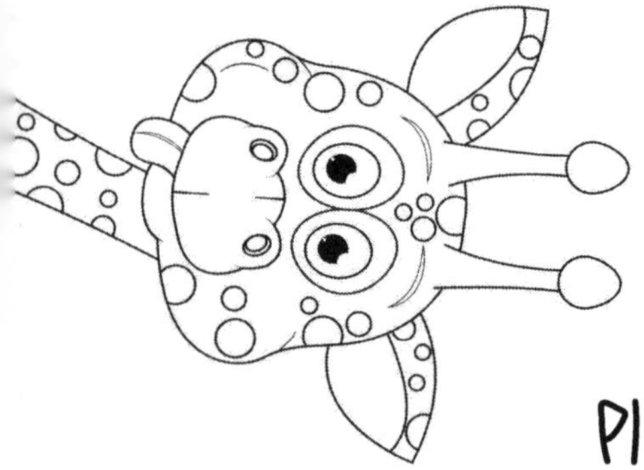

Please leave a star rating or a review on Amazon! Your opinion is important to us. We want you to be happy with our products!

You could... Inspire other kids by uploading a photo of your best work!

"AwesomeKids" (one word)

AwesomeKids Press

☆ ☆ ☆ ☆ ☆

Kid's star rating goes here!

Attention Kids and Adults!

Have Fun In Here!

Graph paper has lots of uses! It is great for learning math, science, writing, making patterns, and drawing shapes.

It is also helpful when learning how to fold 3-D objects like boxes and... paper rockets and paper airplanes! Seeing what the grid looks like on the folded object helps with spatial recognition. It's physics! This may sound a bit advanced but you will be surprised how it helps the brain develop!

We wanted to make this a little more than just plain graph paper so we put a small picture on each page. There is an object on the left page and a character on the right page. This is to encourage the thought process that leads to one of the most exciting uses of math out there! APPLIED SCIENCES! When kids learn how to do math it is amazing. A whole new language! When they learn about it **and** think about a real-life situation where math is actually happening, it makes math even more special, useful, and lots of fun! Using the picture prompts to spark their imagination, they can write math word problems, draw mathematical pictures, color them, etc.

The picture prompts will also help to develop a very important part of the brain, the photographic memory center! They may remember that they drew a really great star and you ask them "Where is that star you drew?" They may answer, "It is on the page with the elephant!"

This is Molly, The Math Monkey. She is all about math and science. She wants to invent outer space vehicles for her and her friends so they can go there to play ________. (you fill in the blank)

I love using this graph paper! I get to use my brain **and** my imagination! High Five x 2!

Note from AwesomeKids Press to the Adults:

In today's digital age, the term 'picture prompts' has taken on a whole new meaning. While our book 'Graph Paper for Kids with Picture Prompts' was created with the intention of using small images on each page to inspire creative thinking about math and science in real-world situations, the concept of picture prompts has expanded to include the use of AI-generated images to inspire writing and storytelling.

As we move forward into a world where technology plays an increasingly important role in our lives, the ability to use prompts, whether they be images or text, to generate ideas and solutions will become a necessary skill for children to learn. With this in mind, we encourage parents and educators to use this book as a tool to not only teach math and science concepts, but also to foster creativity, problem-solving, and critical thinking skills in children.

Thank you for choosing 'Graph Paper for Kids with Picture Prompts,' and we hope that it will be a valuable addition to your child's learning experience.

This is the Amazon Review Link

We would love to hear about your experience with this book!
The future is here! Let's be ready for it!

–AwesomeKids Press

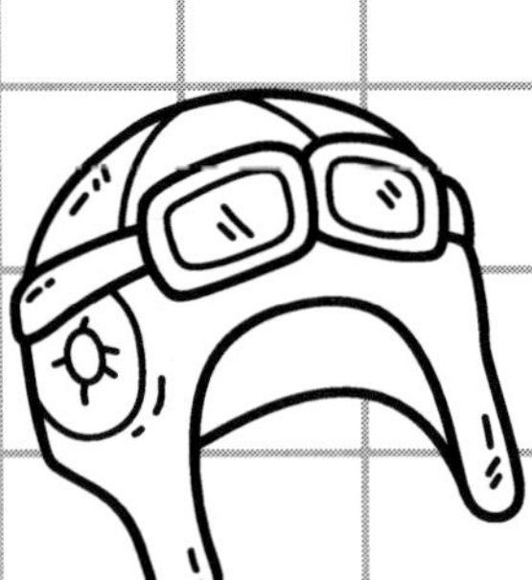

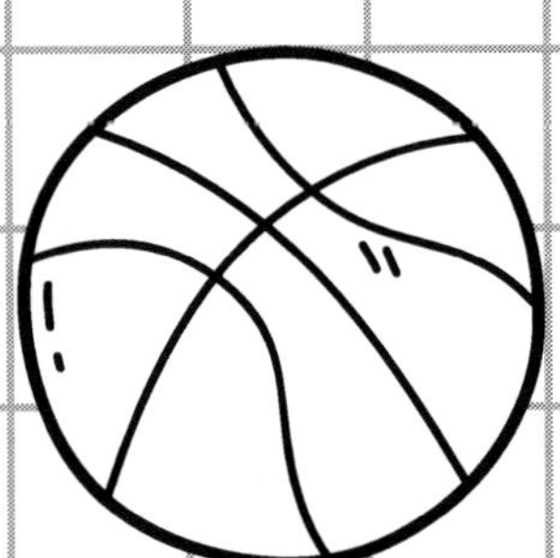

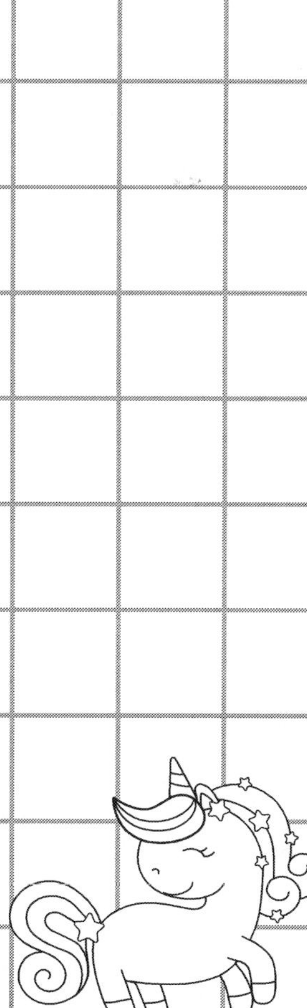

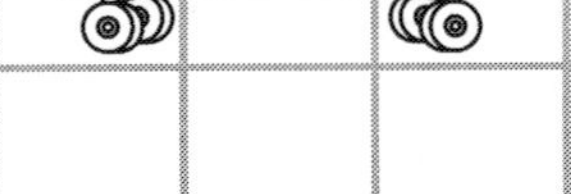

HONEY

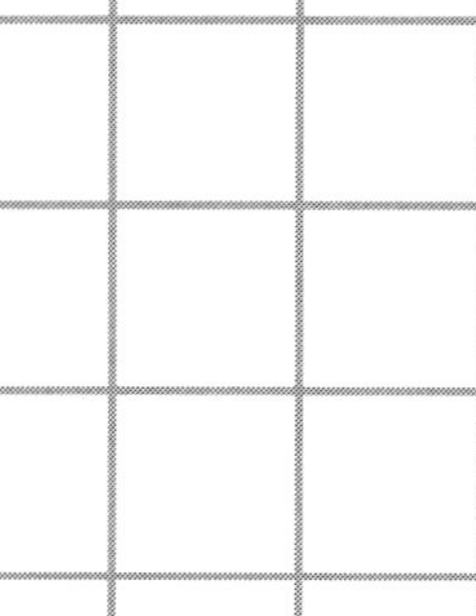

It looks like that bird wants a helicopter too! Interesting! She would have wings AND propellers....

MAIL

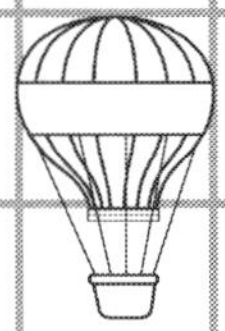

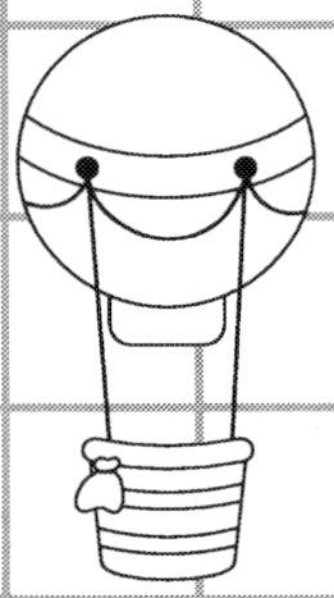

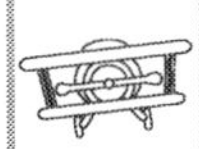

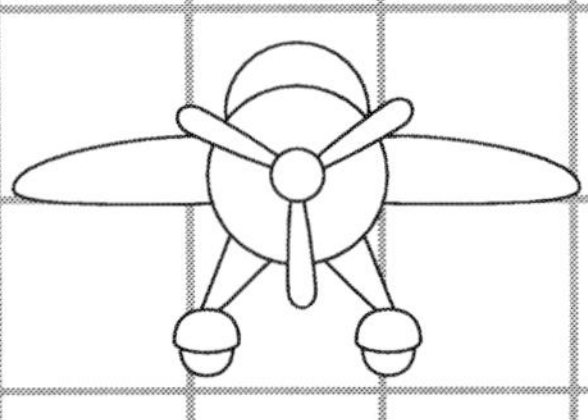

Middle of the Book

1

Plan your vehicle invention...

Make up a new way to travel. It can be anything! I once imagined a bunny who made a boat out of a pumpkin and was wearing an upside-down ice cream cone for a hat! At first, I thought he had 3 ears! He said, "this is no ordinary upside-down ice cream cone hat. It is also my GPS to help me navigate around this magical pond!"

How does YOUR vehicle invention travel? (circle what it will have)

- Does it Fly?
- Does it float?
- Does it dig?
- Does it roll?

Brainstorm

Draw or Write more ideas about your magnificent invention here!

Does it have...

- Wings? Propellers?
- Wheels? Gears?
- Windows? Holes?
- Doors? Snorkels?
- Electronics?
- Solar Power?

Does it have...

- colors?
- patterns?
- what is it made of?

Out of this world ideas!

Remember to turn on your "IMAGINATION STATION"
You can create something extraordinary!

Have Fun!
From Molly
(The Math Monkey)

Invention Activity

2

Draw your invention & give it a name!

The name of my invention is

Draw your invention in a cool place doing something awesome!

MY INVENTION ADVENTURE

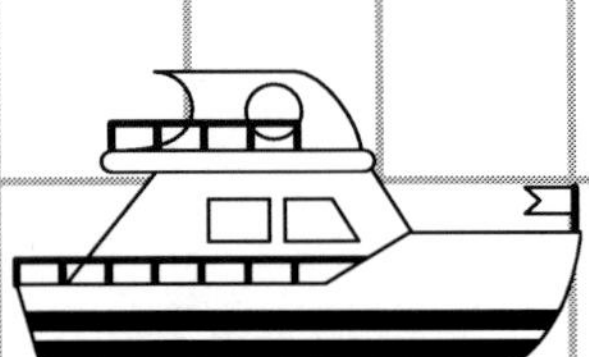

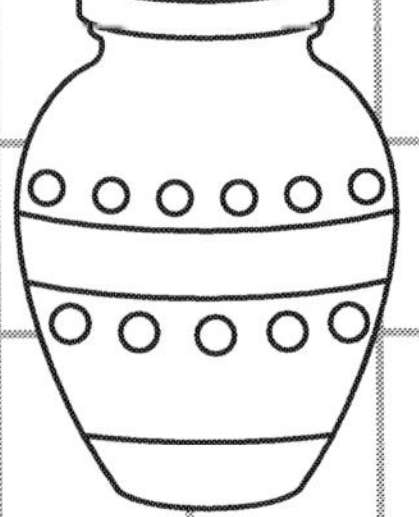

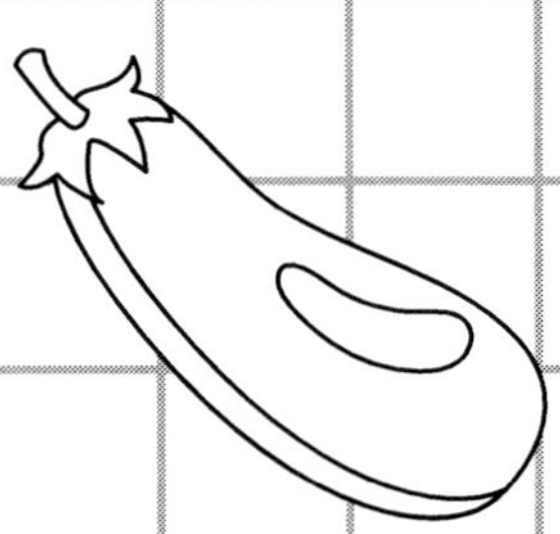

Franco's Fun Facts

Sweet!

My Strawberry Short-Report

- There are about 200 seeds on the outside of every strawberry
- Each little seed is a fruit with an even smaller seed inside! WOW! multiple fruits!
- Does that mean I am eating 200 mini-fruits every time I eat a strawberry? GREAT!
- We find strawberries in the fruit section at the store. They are very nutritious, just like fruit... but the plant is actually a member of the ROSE family! Maybe that is why they smell so good!

By Franco, The Fun Facts Fox!

FUN FACT

Strawberries are the only fruit with seeds on the outside! Well, kind of...
Read 'My Strawberry Short-Report'

This is **'FRANCO, the FUN FACTS FOX.'**
(Try saying that 5 times fast)!
Whenever you see him in one of our books, he will tell you something he has learned from his Fun Facts Research Project.

Hi

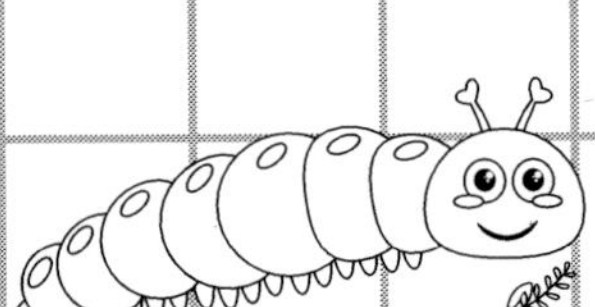

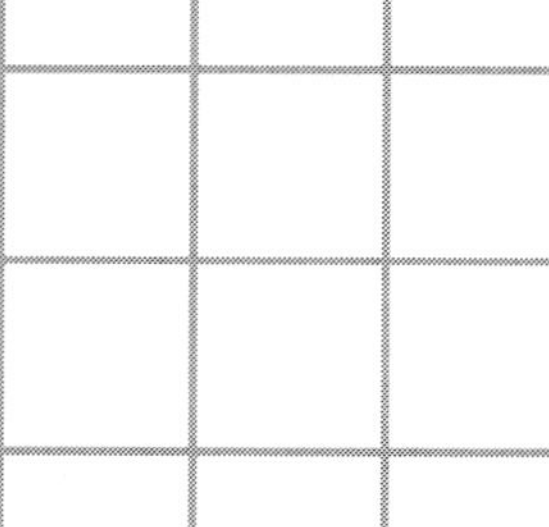

AwesomKids Press

LIBRARY PASS

(on Amazon)

AwesomeKids Press Amazon Library

www.amazon.com/author/awesomekids_press

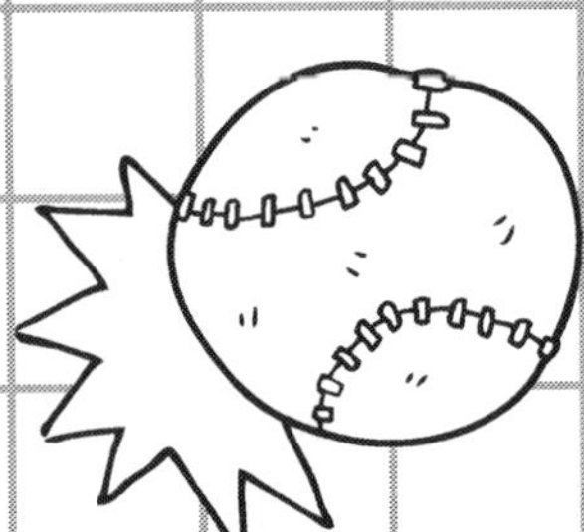

1ST

Nice Job!
Hope to see you
BEAR-Y soon!

Have you ever seen a pink-striped zebra? Zilch? Zero? Me neither! I made some BRIGHT PINK-LINED GRAPH PAPER tho! It is AWESOME!

If you want the

FREE PRINTABLE PDF (8.5"X11")

Ask an adult to send an email titled,

"Pink 1/2 inch graph paper"

To:
animals@awesomekidspress.com

(It's a real email, not a robot one)
(We are really here on the other side)

This is Zip, the Zebra! He works in the Lines Department at AwesomeKids Press! He made the lines in this book!

Made in the USA
Monee, IL
14 September 2026

62980096R00068